Reflections on the Sinister Universe

Greg Feild

May 12, 2017

a feild theory   :)

Abstract:

In this paper, we finish laying the foundation for the universal model of physics.

In particular, we provide a theory of the muon and the three lepton families in terms of the group SU(3).

physics is fun!

Novum Organum:

    Man, being the servant and interpreter of nature, can do and understand so much and so much only as he has observed in fact or in thought of the course of nature: beyond this he neither knows anything nor can do anything.

    The cause and root of nearly all evil in the sciences is this -- that while we falsely admire and extol the powers of the human mind we neglect to seek for its true helps.

    The subtlety of nature is greater many times over than the subtlety of the senses and understanding; so that all those specious meditations, speculations, and glosses in which men indulge are quite from the purpose, only there is no one by to observe it.

-- Francis Bacon

1620

let's do physics!

The domain haunted world:

Gravity was the first force.

Gravitational fields were the first fields. They were not considered to be 'real'.

The electrostatic force was modelled on the gravitational force, and it was found to obey the same 1/R^2 law.

These forces were described by fields and potentials. These fields and potentials were considered to be mathematical conveniences.

They were successful *models*.

Somewhere along the way, gravity ceased to be a force at all, while electromagnetic fields were elevated to the status of real physical things!

Real things existing in space and time independent of the human imagination.

We no longer give credence to these ideas. Only particles and their interactions are real.

In deriving the universal model, we returned to our roots, and based our reasoning about gravity on the successful models of electromagnetism.

The result was a grand unification and the 'final theory of everything' !

There -- we said it!

:)

We have come full circle.

Lorentz force redux:

Let's begin with a closer look at the equation for the the electromagnetic terms of the classical Lorentz force as derived in "On Parity and Isospin".

The electromagnetic force between two identical electrons is

$$F_{1,2} = (e/m_e)^2 (c/R)^2 (\mu/4\pi)(p_1^2 p_2^2/4E_1E_2 \; \mathbf{r} + (1/c^2)p_1p_2\sin\theta\cos\phi \; \mathbf{n}) \qquad (1)$$

where $\mathbf{r}$ and $\mathbf{n}$ are unit vectors. This equation is nonrelativistic. The relativistic form is actually easier to write, and more general, but it doesn't seem quite as illustrative …

$$\mathbf{F} = (e/m_e)^2(c/R)^2(\mu/4\pi)(m_1m_2\mathbf{r} + (1/c^2)(\mathbf{p1xp2xr})) \qquad (2)$$

Equation (2) actually demonstrates the *equivalence* of inertial and gravitational mass!

The (nonrelativistic) equation for the gravitational terms is exactly the same, except for the coupling constant;

$$F_{1,2} = (G)(c/R)^2(p_1^2 p_2^2/4E_1E_2 \; \mathbf{r} + (1/c^2)p_1p_2\sin\theta\cos\phi \; \mathbf{n}) \qquad (3)$$

We note the factor of $(1/c)^2$ dampens the 'magnetic' terms in both equations.

Notice that the factor of epsilon_0 is gone, and we are left with two coupling constants, G and e/m_rest, and a 'space factor', $\mu/4\pi$. The evolution of the force is totally described by the energy and momentum of the two electrons. The factor $(c/R)^2$ describes the time dependence of the interaction, and the time t is common to both electrons.

The particles act on one another. There is no 'source' charge.

This description contains no electric or magnetic fields!

All the energy and momentum of the system is carried by the particles.

Nowadays, we characterize this interaction as two electrons trading energy and momentum by virtual photon exchange. Exchange is the key word. There is no emitter and no receiver; just a continual flux of virtual photons between the two.

Actually, a picture I like better, is of one of a continuous virtual photon, constantly changing 'frequency' or 'mass' as the interaction evolves.

The Born approximation:

In the Born approximation, one assumes the scattering interaction between two particles can be described as the exchange of one virtual photon. This photon is considered to be the most energetic of all the photons exchanged during the interaction, and the contributions from "higher order" photons are found to be negligible in cross section calculations.

Hard scattering occurs when the virtual photon exchanged between the two particles is highly energetic. The frequency of a "hard scatter" in an experiment depends on the incident particle beam densities and individual particle impact parameters.

In the single particle exchange picture, one particle is the emitter, and the other is the receiver. This is an excellent mathematical representation. Physically, it is lacking as the exchange is not 'symmetric', and it is hard to understand how such an exchange could lead to an attractive force.

In reality, the two particles have been interacting and exchanging virtual photons long before they were even collimated into colliding beams! However, we needn't go that far back in time.

The point is, in their final approach down the straightaway, the two particles are interacting furiously. The particles are always interacting; before, during, and after the 'scatter'.

Energy and momentum are always being exchanged between the two particles. There is no emitter and no receiver.

For every action there is an equal and opposite reaction.

The world from scratch:

With hindsight, let's put our new model together one particle, and one particle property at a time, and see what we can learn!

We will start with a single spinless neutrino. It just sits there. We notice it creates a gravitational field that falls as $1/R^2$. Pretty boring.

Next, we give our neutrino some initial velocity. Now, we notice the neutrino creates a magnetic field as well. Interesting, but still pretty boring with nothing to interact with except for our Observer!

We add a second neutrino and note that the two neutrinos can either form the bound state neutrinium, or scatter. If the neutrinos scatter, we notice bremsstrahlung photons and we have discovered our gauge boson!

Classically, the interaction between the two neutrinos is described by equation (3).

We also note that one is able to 'transform away' the magnetic field of one single particle. However, one can not transform away the magnetic effects between two particles in relative motion.

Another reason not to take fields to be real physical entities. Another reason to believe in time and motion.

We also notice the magnetic forces do no work on the particles.   What is up with that? Why would nature create a force that does not change the energy state of the particles during an interaction?

We realize that particles must have a spin that interacts with the magnetic field!

Now, with just one kind of particle (and antiparticle), and one attractive force, we still can't build much of a world; even with spin!

So, we add the electron. An exact copy of the neutrino, but more massive, and with an extra, and significantly larger charge, that can be both positive and negative.

We now have all the raw materials to build a universe!

The speed of light:

Why can't a particle travel faster than the speed of light?

I believe the popular response is because the mass of a particle tends to infinity as the velocity of the particle approaches the speed of light.

This is true, but incorrect.

A particle cannot exceed the speed of light because it cannot exceed the speed of the force (or the source) causing it to accelerate!

Classically, we can imagine placing an electron in a constant electric field of infinite extent. This field would exert a constant force on the electron and, in principle, we could accelerate the electron to any speed we'd like.

However, fields are not real.

Our electron's acceleration is actually affected by the exchange of virtual photons with a capacitor plate.

The faster and farther our electron is accelerated, the farther the next virtual photons have 'to travel' to give our electron its next boost.

Time and space:

Time and space are emergent (to use the popular parlance) properties of the relative motion between objects. Time and space arise from our need to assign positions and velocities to objects in our study of dynamics.

No objects; no space. No motion; no need for time!

Know objects; know space. (It had to be done!)

In any given inertial reference frame, I believe it is appropriate to define time and space as Newtonian, at least operationally. We have a fixed three dimensional space and a ticking clock.

The only new feature we impose is that a particle's mass is now dependent on its velocity. However, this is not a comment on the nature of time and space.

Particles 'lose and gain' 'kinetic energy' during an interaction and this is reflected in the particle's varying mass-energy.

We work with the Lorentz transformations and write our four vectors in Minkowski space because a particle's mass-energy is a function of the the (relative) velocity of the particle.

Is time eternal? Yes. Is that tautological? Yes.

Is the universe infinite in extent? That is impossible to say. Perhaps, one day we could possibly discern the edges; otherwise …

It is really too early to say anything about the universe until the current cosmological data is analyzed in the light of our new model.   The universe may actually be collapsing right now!

But, probably not.

Matchbook summary:

The universal Lagrangian is the QED Lagrangian; except that we need to introduce the mass dependence of the charges using the energy operator

$$E \ = \ - ihbar\partial/\partial t \qquad\qquad (4)$$

The exact gravitational and electromagnetic Interaction terms in the universal Lagrangian are;

$$\text{L\_interaction} \ = \ \text{-i*hbar}(G - e/m_{rest}) (\psi^{bar} \gamma^{\mu} A_{\mu} \ \partial\psi/\partial t) \qquad\qquad (5)$$

As usual, we are not really confident about the sign convention, but equation (5) should be correct for either the electron or the positron!

At least there are no pesky factors of pi.

Group theory:

The group SU(3) reflects an exact symmetry of the universal model, operating on both the charged lepton basis

e   = (1,0,0)

muon = (0,1,0)

tau   = (0,0,1)

and the neutrino basis

nu_e   = (1,0,0)

nu_muon = (0,1,0)

nu_tau   = (0,0,1)

The neutrinos have the same mass hierarchy (~1,100,1000) as the charged leptons.

The eigenvalues, q, of the charge operator are degenerate for the charged lepton basis, but the eigenvalues q/m are non-degenerate, as are the masses themselves.

We can use the step up and step down operators (I+ and I-), composed of the Lambda_i matrices of the standard model, to transform between the particles in each of the two independent bases. Neutrino mixing, anyone?

We can also borrow the formalism of flavor SU(3) by making the substitutions

up → electron

down → muon

strange → tau

and generate all the pseudoscalar mesons, for example;

pi^0 = 1/(2)^½ (e e^bar - mu mu^bar)

Finally, we can generate particles by mixing the charged lepton and neutral lepton bases, and the neutral lepton bases, using the Lambda_i.

Conclusion:

The world is nothing but mass in motion, and *all mass* is in motion.

Even a particle "at rest", is a mass spinning to the left or to the right.

As we know, all massive particles experience the universal, mutual, gravitational attraction which is proportional to their masses and separation.

In addition, there is an interaction between particles that arises due to both their relative and absolute motions.

Just as two parallel (wires carrying) currents of opposite charge will attract one another, so will two massive particles spinning in opposite directions (i.e. electrons and positrons), leading to their complete and total annihilation!

Well, that's about all I know!

I turn it over to the universe.

Gravity:

The weakest force, the strongest force, and the 'weak' force!

About the author:

Greg Feild is dark
and mysterious!

He lives in the present.

References:

Classical Dynamics of Particles and Systems
Jerry B. Marion

Foundations of Electromagnetic Theory
John R. Reitz, Frederick J. Milford, Robert W. Christy

Quantum Physics
Rolf G. Winter

Gauge Theories in Particle Physics
I. J. R. Aitchison and A. J. G. Hey

Quarks and Leptons: An Introductory Course in Modern Particle Physics
Francis Halzen, Alan D. Martin

Quantum Field Theory
F. Mandl, G. Shaw

Symmetries and Group Theory in Particle Physics
Giovanni Costa, Gianluigi Fogli

Quantum Mechanics: The Theoretical Minimum
Leonard Susskind, Art Friedman

just another seven dollar theory

Books by Greg Feild:

1. "A quantum mechanical theory of gravitational interactions"
   CreateSpace Independent Publishing, 8/29/2016

2. "Observations on the quantum mechanical nature of gravity"
   CreateSpace Independent Publishing, 10/8/2016

3. "On gravitation and electric charge"
   CreateSpace Independent Publishing, 11/1/2016

4. "On spin, mass, and charge"
   CreateSpace Independent Publishing, 11/29/2016

5. "On angular momentum, acceleration, and absolute motion"
   CreateSpace Independent Publishing, 1/4/2017

6. "The Sinister Universe"
   CreateSpace Independent Publishing, 3/1/2017

7. "On Parity and Isospin"
   CreateSpace Independent Publishing, 4/11/2017

the heptateuch   :)

Appearance is not reality:

Reality -- What a concept!

Notes:

Check out your mind -- The Impressions:

Been with you all the time!

www.ingramcontent.com/pod-product-compliance
Lightning Source LLC
Chambersburg PA
CBHW081136180526
45170CB00008B/3120